Raising Interesting People
Collection #1
<u>Why Send Them To School?</u>
And other essays by
Meredith Olson Ph.D.

Better known by her students as
Doc "O"

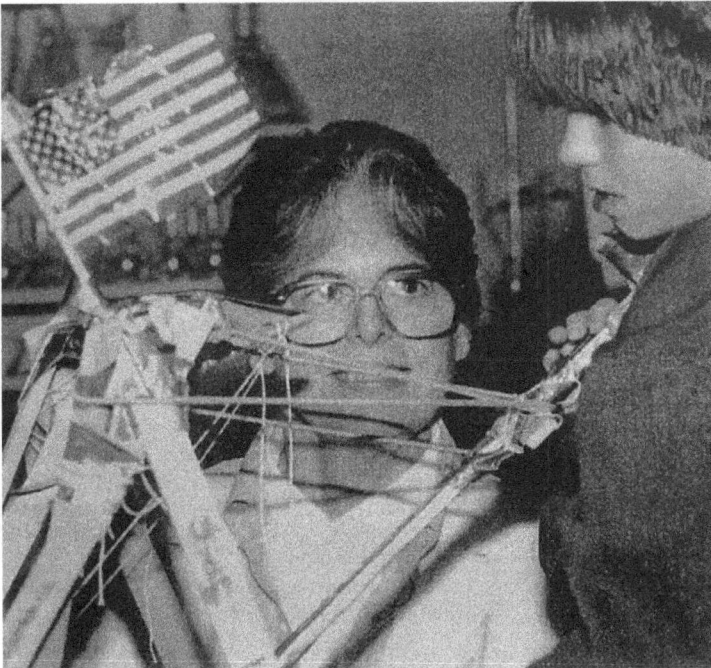

<u>www.DocOsBooks.com</u>

Published by:
SAN 299-2701
Glenhaven Publications
4262 NE 125 Street
Seattle, Washington 98125

        ISBN   978-0-9657061-3-1 (paperback)

Table of Contents

Preface

My neighbors have begun to take an interest. For years we just came home and exchanged pleasantries, "How are you." Recently that has changed a bit. I've been asked to describe the engineering event. I've been asked to explain what went on today. As a result, I have begun to jot notes on the events of the day. When the little essays are read, I am pummeled with questions. "Why did you do that?" "I never thought of that in that way!"

It is hard to explain what I do, yet the kids love it. They express great dismay when they have to miss a science class due to sickness or family trip. There is something wonderful going on – but how to characterize it?

And who would want to read about the events of my day? Perhaps the words I use will raise a memory in my former students. These are the words I use in my classes every day. These are the thoughts that go through my head as I try to penetrate the soul of each child. So, as you read the essays, remember the day ....?

Taken together, they describe a deep and abiding philosophy of teaching and a respect for the mind of the child. I am told that I teach differently than others. Perhaps reading these little vignettes will spark reflection from other practicing teachers. I am passionate about what I do. I work very hard to make it coherent and consistent. I love to learn and am privileged to work at a place that allows me intellectual freedom. I hope my energy and enthusiasm rubs off on the kids. It often seems to.

So, here are some daily musings about whatever comes to mind. The readers will have to "catch the idea" and make a theory of it for themselves.

Interesting People

Interesting people have the capability to talk about interesting things. Will our children grow to be like that? Why do we have children? Why do we spend so much on them? Time? Money? Is it because we have to? Maybe they weren't planned, but once they are here, they are amazing to watch. We become attached to them. But what do we want for them? What do we hope the outcome of our effort will be?

Interesting people have the capacity to be engaged. But you don't develop that capacity by talking about it. You develop it by experiencing it.

To have discourse you have to listen, really listen. You have to go beyond words to the visual imagery and cognitive constructs behind those words. The other day I asked my class how houses burn down. I described gigantic Yule logs in medieval England and suggested that they may have warmed houses for months in the winter. I waved my hands in the air encompassing a large campfire of logs and then cupped my hands over the size ashes that were left when the campfire was over. Then I asked the class where the logs went. How did it happen that the ashes were so much smaller? One student (Martin) explained that there are many tiny tubes in trees so actually there is very little fiber in trees. He said trees are mostly air. He said that when trees burned the small amount of solid matter broke apart and became ashes and the smoke given off was simply contained in the wood beforehand. Now, I know Martin has been exposed to the equation for photosynthesis as I had presented it to the class a few months ago. It seems that he assumed that the carbon dioxide and water taken in actually became the tree. He knew that carbohydrate and oxygen were given off but it is obvious that his estimation of the mechanisms were way off. Most significantly, he showed absolutely no awareness of chemical change. The air stayed air. Fascinating!

To promote discourse, you have to leave non-closure. You need to leave ideas dangling. I told him that his idea was new and interesting to me. I also said that I had a different way of thinking about it. Although I am the teacher, students are used to me tricking them and confronting them with new questions. With a smile, I told him that I intended to do Brain Surgery on him. I was going to try to change his brain – the neural connections in his brain. By that I meant as Marion Diamond had determined from her work with rats, I was going to give him more experiences so he could compare and contrast new data with his current line of reasoning.

Then I reminded him (and the class) that it is OK to change, modify, revise ideas. I said "Think of a 4-year old who believes in the tooth fairy. You no longer believe. Are you bad? Was the 4-year old bad for that belief? No – ideas just grow and change and that is OK. It makes life interesting."

So, notice: authentic discourse must be humiliation free. It must be OK to seek new views and to hold on to old as long as we want to. No grades or judgments are attached to the development of these wonderful ideas. Each idea has merit, but then we move on.

Interesting people are not intellectually rigid. They are kind and contemplative. They are able to actually listen to others. They love putting ideas together in new and intriguing ways. They like to follow ideas to their logical consequences.

Interesting people are interested people. How do we help them become so?

Why Send Them To School?

Our emotional maturity and cognitive curiosity are, in large measure, derived from early experiences of teacher-student and student-student interactions. As depicted in Lord of the Flies, students may not be able to structure a generative environment for themselves. So, what does a teacher add to life's experiences? What actually happens in school? What kinds of cognitive habits are formed in the classroom setting? Is the classroom dull or interesting to the child? Does classroom activity develop interesting people?

Is there a difference between being competent and being interesting? Do schools strive for competency? What should be happening at school? Are schools "information givers"? Is the web providing new avenues to gain competency? If we can become competent by watching the History Channel and by viewing math learning projects, then why go to school? Filling in a bubble on an answer sheet may make you feel competent but it does not produce reasoned arguments, serendipity, or satisfying moments of "Aha".

What should we see in classrooms actively engaged in developing interesting people?
Teachers who pose substantive ideas for discussion and then sensitively listen to student-student dialogue can guide them to habits of mind which will enlarge over a lifetime. Classroom protocols can enhance the likelihood that students will explain their ideas in public, free from ridicule and censure. Students can learn to hold multiple working hypotheses in mind as they "play with" new experiences and information seeking new

connections between ideas. An attentive teacher who listens actively can come to understand the constructs the student holds and thus can provide experiences and explorations which deeply interest their students.

But why school? Why not mentor-student, parent-child, or wonderfully programmed individualized computer instruction? Each of these relations has merit. The added value of a group of students is that they may be at similar cognitive levels and so may be better able to probe the logic of the other. They provide practice in interesting discourse among peers. They may develop lifelong habits of talking about interesting things. I am always pleased to overhear students in the hallway heatedly discussing issues raised in my class debate. Amazingly, they are pursuing science debate simply because it is fun. This suggests a developing ability to talk about interesting things on their own. Just today, after having students closely examine burning candles (in pairs of students) at lab, three young boys swooped in after school wanting to make more comments. I tape record each class I teach and students often strive to get their reasoning on tape. The three boys asked for the recorder to be turned on so they could document the reasoning they had been doing together all afternoon.

Their ideas were not complete, nor even quite correct, but the engagement was certainly interesting.

The Airplane Takes Off

Bronson was absent from my class yesterday. He arrived in class on time today and spent the pre-class break amiably talking to classmates. When class started he loudly announced, "I was absent yesterday. What did I miss?" Being in the midst of a unit about the structure and function of the human heart, we had discussed the value of the Foramen Ovale and Ductus Arteriosus before birth. I had posed questions about whether there was a need for already oxygenated blood to go to lungs before birth. We had puzzled and marveled over ways the blood shunted to the quickly growing brain and the rest of the body.

Then birth! The drastic environmental crisis. How did the baby cope? How could the whole system transform in a matter of minutes from umbilical oxygenation to alveolar oxygenation? Blood had to go to the lungs and do it quickly. No time to slowly grow. How did the remarkable change happen? The magic of the story wove itself into the minds of students via teacher questioning and students boldly grappling with possibilities. Imaginative solutions were embraced. The Ductus must become a ligament just as the umbilicus does. The Foramen Ovale must have a flap large enough to close the opening when pressure in the heart chambers changed.

But what if the flap was too small? What if the Ductus was too big? What had been useful before birth was now a life crisis. PFO. PDA. The baby was blue and gasping for oxygen. What could a Doctor do? We examined stents and occlusion devices and played with catheters which might insert them. We pondered the effects of one procedure and then another. We felt the pathos and exhilaration of striving to improve a perplexing situation when we don't know what to do. And, at last we recognized that doctors are fallible humans just like us. They struggle to save a life and they often succeed. At times the challenge is too great and the baby dies.

Drama. Engagement. Problem discovery. Problem solving. Success and failure and struggle and intellect. Intense, interesting conversation.

When you go to the airport, you must be on time for your flight. If you are late, the plane takes off without you. Bronson can look up the medical jargon on the web, but the irrepressible enthusiasm of class interaction is gone. Flown away. If you miss class, you miss a magical moment … forever.

Never Trust a Science Teacher

Repeat after me: "All science is based on a sugar cube." Now, after writing up the lab, go measure twelve items in the room using a sugar cube. At the end of the activity you may trade in the dirty cube for a new one. Finally, dispose of the cube properly.

The next day students share their measurements in class discussion. After every student has contributed a measurement, I say, "I lied." All science is not based on a sugar cube. At least I sort of lied. All science is based on a smaller plastic cube like this one (which I hold up) but it doesn't taste as good. The students do not know it yet but, over the years, they will learn that a cubic centimeter is the basic measurement for a gram, a milliliter, a calorie, and even the speed with which continents move about on earth. Science is indeed based on a cubic centimeter!

So I say, "Never trust a science teacher." I am always trying to catch you off guard. I am trying to lead you a bit astray. Can you catch me? Can you notice when I am wrong? Heisenberg and others established various Uncertainty Principles. They found one could not precisely measure position and momentum at the same time. As one measurement became more exact, the other became less exact. Now, a sugar cube is not precisely a cubic centimeter, but it tastes good in a memorable way. We now can conjure up an image of the size we are talking about in future studies by remembering that: "All science is based on a (small, not so tasty) cube." Each year when I get to this lesson, students throughout the school know they can visit my room to retrieve their yearly sugar cube if they repeat the words to me: "All science is based on a sugar cube." Silly. Playful. But it marks a spatial memory in the mind of a child.

So now we need containers. We fold foil around plastic centimeter cubes. Some leak. Some are smushed. "Who do you think my hero is?" I ask. Students typically suggest Einstein. "No, Mary Poppins!" She was practically perfect in every way. Our pots are not perfect, but they are practically perfect pots. Even though not quite perfect, we will use them as containers in several future experiments. We can't be perfect in science, but we can be close. We have uncontrolled variables in everything we do.

So, even though we try to take careful data, we never quite believe it. We graph our findings and look for regression lines. We learn to work with an imperfect interpretation of the world – and we never trust a science teacher.

Two Divided By Two-Percent

The students were told that when they finished their test they could read, draw, or work on homework. One by one they handed in their papers and began to work on something else. Dan commented, "I'm on my 6's." "You are?" "Oh my goodness! I'm not finished with my 4's."

Students had been assigned an amazing problem in math class. The task was to write the value of all the numbers from one to one-hundred using four 4's – no more and no less. They could use any math signs they wanted to use. Well now, lets see. 44/44 is one. 4/4 plus 4/4 is two. The square root of 4 times 44 plus 4 is 92. OK, this works. Students knew they could give each other hints but not actually work problems for them. They could offer information about new mathematical signs such as factorials. (4!=4x3x2x1)
After cracking about half of the number values from one to one-hundred, students discovered the value of dividing by decimals. Four divided by .4 equals ten. Quite useful. But four divided by .4 repetant equals nine! Amazing! Soon students were checking others work to detect errors. Many said, "I have just three left to do." With joy and satisfaction students, one by one, achieved solutions to all one-hundred problems. Celebration for doing math problems! No grades. No gold stars. Just satisfaction. How unusual!

Some students went on to try to represent all one-hundred values using five 5's. That was much easier. Dan had completed that. Liking to be first and best, he moved on to using six 6'. He liked being ahead of the others.

Students were discussing their clever inventions with others when, across the room, Dan spoke out. "I can make 100 using two 2's."

The class fell silent. Stunned. Thoughtful. Finally I said. "Ok, how?" Dan replied, "Two divided by two-percent." Laughter. "Oh my goodness!" "That really works!"
Students were delighted! What a wonderful invention!

You see, students had to work on the problem. Become invested with the problem. Become puzzled by the problem. Get to know how mathematical short-hands might be of help. Finally, after enough familiarity with the problem they were able to recognize and celebrate.

Interested people discussing interesting ideas.

Witness! Witness!

Children compete. They spend much of their time in school searching for their place in a group.   Who will they sit with? Who will let them join? Who is popular? Who leads? Can competitiveness be harnessed to promote excellence instead of divisiveness?

We spend five weeks each fall developing an engineering event. We choose a different theme each year sparked by some news event. This year's theme was Wind Farms. Students typically each build four or five mechanical devices in compliance with specific design criteria. The first creation is usually some sort of "go" machine which moves across the floor. As my room has one-foot floor tiles, distance traveled is easy to measure. Using junk, tape and hot glue, students attach wheels and axles to some sort of chassis according to the design criteria specified that year. There are many technical difficulties to be worked out. Axles need to be aligned. Bushings and bearings have to be created. Students have to discover that it is not helpful to glue the axle to the chassis. The motive force chosen for the year has to be housed and engaged to make the vehicle move. Every student makes a rig.

As soon as someone gets their vehicle moving, their name is put on the board. Beside their name is listed how far the vehicle went in floor tiles. As more students get vehicles moving, more names go up. The list grows until every child is on the list. Each time a child increases their distance moved, a new record is added next to their name. Students can watch their distance improving and compare it with distances moved by others. It is hard to keep track of start and finish positions, so every new record must be verified by another class member willing to act as a witness. Cries of "witness, witness" ring out as students request verification of distance in a run they are about to make. Having a witness promotes conversations. "Wow! How did you do that?" "I want to align my wheels and do it again." Soon students are looking, really looking, at other's rigs to discover what works and to get ideas to improve their own rigs.

Our events have no prizes for winning or even for improving. Every child who enters gets an entry prize – and every child must enter. The entry prizes are little candy models made according to the year's design. This year's prize was a chocolate-kiss and lollypop anemometer that actually moved when you blew on it.

So – no prize for winning – but everyone who tries gets a wonderful working entry prize. Students take notes to record how well their own rigs performed. Success and the feeling of accomplishment are their own reward.

The having of witnesses generates conversations and sharpens attention on details of mechanical constructions. There is just enough competition (recognition of self improvement and celebration of the improvement of others) to keep enthusiasm high.

Witnessing draws us together and initiates conversations about devices in which we are all interested. Interesting conversations. Interesting people.

Class Size Matters

The web is full of educational plans for big classes. Charter schools advocate computer assisted learning which, they say, provides individual attention for each person in a large class. I suppose the frequently found teacher lecture style presentation can address a class of any size. So, perhaps the only limitation is the availability of restrooms. Students certainly seek them out more often when classes are isolated or regimented. But what determines "my" class size? I know that class flows best with ten to fourteen students. Magic happens then. Why is that?

More that sixteen students poses a problem. With only forty-three minutes available for each class (and I don't want double periods – the time is just right to promote incubation over night), I have to plan the class flow. Perhaps I will take five minutes to set up the question. Students will need about twenty minutes at lab. That leaves about fifteen minutes for class dialogue. Now, if every child speaks once, they have less than a minute to present their idea. With class data reporting and reasoned reflection, each student should be commenting two or three times. Thus each comment can only be 20 seconds long. Every time I add another student, the time available per child gets shorter. I am almost forced to ask short answer, less engaging questions.

Now suppose I have 20 students. There was time when I did – and the parents went up-in-arms over it. What actually happens is that about ten children participate in lively debate. The rest listen pensively and take notes. More students: more quiet observers. The more students in a room, the lower the percentage of actively involved students.

But now, lets consider lab. If one student does the lab for a teacher, they have to get the answer "right." The only comparison is the textbook. If class contains ten students, there are five labs to collect data. (I find the number of students per lab makes a huge difference in their logical discourse and ownership of the data. Two is perfect.) Now five labs can generate enough data points to create a regression line when data are reported to the class. Eight labs make a nice amount of data but even middle school children loose patience with data overload. I can just barely keep their enthusiasm going with "Oo's" and "AH's" and, "That fits the pattern nicely!" With more than eight data sets, we run out of time.

Even the roughest, sloppiest lab can generate generalizable data. If you don't average it you can see the trends. With enough replications, some labs get data close enough for conversation. I think of a particularly sloppy lab asking about the possible arrangement of arteries and veins in your legs. Pouring water into copper or plastic leg-length tubes allows temperature data about blood exiting the body core and then entering feet. Warm water from body. Cold water from feet. We run flows though separate tubes, parallel tubes with concurrent flow and then parallel tubes with countercurrent flow. Though we use standard laboratory thermometers, foam cups, and water from the sink, we can actually develop the concept of countercurrent heat exchange. That only is possible if you have just the right number of labs. Otherwise, I would just have to tell them.

And telling doesn't allow skepticism, questioning, and openness.

Do Candles Burn?

Who can burn a candle fastest? Students were given 12cm white camping candles. They scratched cm marks in them so they could reference how much candle was left after each minute of flame. They held them in wire test-tub holders over foil trays of water to catch any possible drips. Each lab pair diligently watched the process for ten to twelve minutes.

The following day each lab pair reported their starting length, ending length and flame time. Some labs had only a centimeter of candle remaining. Then I asked everyone to go to lab and approximate how much wax was in the water tray. Students proudly reported enormous amounts of collected wax. Then I spoke with alarm. "You what?" "You melted the candle?" "You were told to burn the candle!" I gave each student a blank paper on which to write what had happened. Without exception, students reported that candles do not burn – they melt. The flame (which had lasted more than ten minutes) was burning the wick, but the wax melted. A few students noted that the amount of wax in the pan was not quite enough to rebuild the entire candle so some wax must have evaporated.

Hummm. OK. That is good progress. Now, what do I, as the teacher, do? In previous years most of the students had gone on a field trip to Pioneer Farms where they were able to use tools from days gone by. I asked them if they remembered the scissors kept near the candle holder used for trimming candle wicks after each burning. If wicks burned, why did they need to be trimmed after each burning? Asking for written response on what three things do flames need to keep going, every students

was able to say fuel, air, and heat. So, what was the fuel? The
wick? How could such a small thing as a wick keep the flame
going more than ten minutes.

Then we examined the shape of the wick. I told them that early
wicks just stuck up above the flame and got in the way. The
wicks we saw curved over at right angles after going up about
one centimeter. When flame existed, we noted the very tip of
the wick had red glowing embers. Amazing! Automatic
scissors! What a clever invention to trim the wick as the candle
burned!

But is the wick burning? How can we know if only the tip of
the wick is burning? We held a toothpick in the top of the
flame and found a singed spot on the toothpick where the flame
had been. Then each lab held a toothpick low in the flame. We
discovered that there were now two singed spots with an
unburned section of toothpick in the middle.
That must mean the flame only "burned" on the outside of the
flame. The interior was not actually burning! How to account
for this? Well, if the flame was all around, and if flame needed
oxygen, maybe no oxygen could get to the center of the flame.
Flame needs three things, but the center of the flame had only
two.

So, if the wick only burns at the tip, what is the fuel for the
flame? We will work on that next week. These students are not
ready to read Michael Faraday's lecture on the nature of a
candle given to children in 1860 – but some day they will be.
Hopefully they will find it interesting.

Worms-With-Clothes-On

Children think fire paints things black. Something is added to
the outside of burned things.   But is it?

Well, actually, burning means ripping clothes off. Worms-with-
clothes-on.

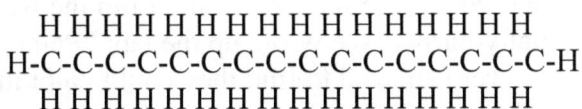

```
      H H H H H H H H H H H H H H H H
      H-C-C-C-C-C-C-C-C-C-C-C-C-C-C-C-H
      H H H H H H H H H H H H H H H H
```

We had drawn the hydro-carbon molecule before. We now
could ask what we would see if the hydrogen clothes were
removed. Why, it would look black because the carbon would
be revealed, of course!

Now, our melting candles project revealed that students believe
that candles don't burn even when flame lasts for ten or twelve
minutes. No, No, candle wax doesn't burn. Wax only melts.
Wicks burn. Wicks are made of plant fiber so we know they are
C&H. Worms-with-clothes-on. We say,"Heat is jumpingness."
Heat the cotton wick. All the atoms should jump around. The
hydrogen "clothes" should jump right off of the jumping bodies
leaving black nakedness. We see smoke. That is carbon. Can
we catch it so we can collect evidence? Can we catch the
"naked bodies" jumping around? Lets hold a flask by it's neck
and rotate the round base in the carbon smoke. Oh my!
Catching naked bodies in science class!

YES, YES! That worked. Students were able to catch small spots of blackness on their flask. They left the room saying, "That was the coolest class!"

Students gathered in class the next day with their soot-stained flasks. They estimated how large a spot of blackness they had grabbed. They had held a flaming string (wick) under the flask. What would happen if we held a candle (wick and wax) under the flask. We think wax doesn't burn, but if we could catch more naked carbon bodies, that would mean that something contributed extra carbon. It must have come from the wax. Maybe wax actually burns! We wrote up the lab: "Paint a flask black." Students were amazed to find they could paint the entire bulb of a Florence flask velvety black. Where did all that blackness come from? The only available source (besides the string) was the candle wax. Wax must actually "burn".
Looking carefully we could see spots of water on the flask. We could also see spots of cooled wax. The molten wax must have evaporated and cooled to form solid wax again.

We look at the equation for photosynthesis again, and at the equation for burning.

$$CO_2 + H_2O \ \text{-----} \ C\&H + O_2$$
$$C\&H + O_2 \ \text{-----} \ CO_2 + H_2O$$

C&H breaks apart and recombines with Oxygen. We get invisible CO_2 gas and water spots. We are right back where we started! What a great cycle. It is beginning to make more sense.

Constructing Paper Cranes

Japan has had terrible earthquakes. (They had another today.) Our kindergarten through fifth grade students tried to help Japan by making Origami Paper Cranes. These Japanese symbols of hope will be sent to Students Rebuild in Seattle and trigger a large monetary donation from a local family foundation. The cranes will be woven into an art installation for Japanese youth from American children.

I was assigned to watch and work in the first-grade classroom. There were several adults and student helpers in the room in addition to the regular teacher, who was in charge. Paper was distributed along with a sheet of pictorial instructions (from the web) and the students began to try to fold cranes. The first-grade students had worked on cranes a bit in art class and several easily made the first four folds of the paper. Soon it became confusing. Little hands pushed on the paper folds rather than creasing it with fingernails. Helpers at the various tables tried to instruct using the linear set of directions that the first-graders were to follow step by step. Both the adults and the children approached the task as a sequential set of memorized steps.

Finally, one little boy (Duncan) crumpled up his paper and, with tears in his eyes, put his head down saying, "This doesn't make any sense." I knelt down beside the table near the boy. I knew that his understanding had to be referenced to experience he had had which was meaningful to him. He had to discover some structure in the paper he was folding.

We looked at a nearly completed crane. I asked, "Where do the 'legs' come from." I tweaked up and down the legs. "What is the first fold where we see these?" We started folding a new crane. He was able to make the diagonal folds and the horizontal and vertical folds. "Now we are ready for the first 'collapse' of the figure." This changes the large square into a smaller square. I said, "Look under there: Peek-a-boo, there are the legs." "Hiding under the top flap." "Can you find them?" So which way is the next fold? Why, it has to double the hidden legs so we strengthen them. After each fold I asked, "Where are the legs now? He found them and corrected a couple of folds which went the wrong way. Building legs made sense. But now we have a problem, "Duck feet." There is a web between the legs. We need to get that web out of there. So we grab the outer layer and fold it up. Ah, that makes wings! And now the legs are free. We turn the crane over and make another raised wing on the other side. Our legs are separate, but they are fat. Fold them over one more time to make them skinny and strong. He folded, flipped it over and folded again. Now we have to transform the legs into a tail and a head. Now fold down the wings and viola: a crane! Duncan beamed. He was radiant! He reached for a new paper and was completely focused on folding a new crane.

I moved on to work with other children. A half hour later Duncan had folded six cranes. Each one was more precisely done than the last. He was totally energized by his new understanding. I had witnessed a transformation of the human spirit – from mindlessly doing what he was told, to experiencing delight in discovering meaningful structure.

So, how do children come to make sense of what they do?   How do we teach?

Don't Say Good Morning

The Principal came to my room this morning. A parent had stopped by last night complaining about me. It seems that their son had said, "Good Morning" to me and I had asked him not to do that.

I remember the event clearly. I was starting class. I had just turned on the overhead projector. Jeremy had leaned over to me saying, "Good morning Doc "O". How was your weekend?" I paused and said, "Please don't do that." Think of an orchestra conductor. When you arrive at the concert, do you go up to the orchestra pit as the music is starting and say to the conductor – "Good morning, How was your weekend?" Jeremy sat down. I guess, from the fact that he told his parents about it that night, he must have been upset. Perhaps confused. Good parents and other teachers take pains to train kids to say good morning. Kids are told that saying "Good morning" is a sign of respect. They should do that. But should they?

If children love a class and a teacher, they come to school with an urgency to communicate. They want to reestablish the bond. Saying "Good morning" reconfirms the relationship. But that brings us back to the purpose of coming to school. Students can talk to parents. Students can talk to tutors. Students can interact with computer learning programs designed to promote their steps in understanding. These are all one-to-one interactions. If the purpose of gathering students into classes is to promote the habit of having deeply interesting dialogue between them, then we need to use every tool we have to do that. The power of teacher attention is remarkable. If I listen, actually listen to you, you become aware that you are worth listening to. One of the most powerful things I, as a teacher, can do is ask a question and then deeply listen to students discuss it. They get my attention when they say meaningful things to each other, and really listen, and reply. The power of a teacher should be used to promote what really matters.

Even coming in before class, as Ella often does, relieves her tension and confirms her relationship. Since I am usually preparing chemicals, she thinks I am free to talk to. What she doesn't know, and I can't tell her, is that I am eves-dropping on students as they enter the room while I am turned way and not looking at them. I am assessing their mood and trying to notice current social emergencies to use as an entry point for the day's lesson. Even when I seem to be available for conversation, I am not.

Saying "Good Morning. How Are You?" in the hallway is just fine we both know we do not plan to stop and talk. We are not starting a real conversation about how I am.

The way we greet defines whether we promote shallow or deeply interesting conversations with children. Students can learn the difference.

Don't Average Data

Each student comes to class as an individual. They are the center of their experience. They are the ones who perceive their human condition. They have personal identity. We crush that identity when we make them part of the crowd. School numbs identity in subtle ways. The class clown, the leader, the bully – all are defining their identity. Kierkegaard reminds us that individuals are responsible for living their own lives, passionately and sincerely. **Why do students hate science?** Students don't know why they resent it, but they do. Can it be that science, the way it is typically taught, violates their search for self? **What is wrong?**

Near the end of last semester, students were running "Hot Rod Races." The challenge was to see who could make heat from an alcohol burner travel from one end of the rod to the other fastest. But, how do you know if the rod is hot? It is against the rules to touch it. No branding in science! The rods were aluminum. They showed no change in color as their temperature changed. How can we devise a way to remotely sense data – to tell when and where the rod was hot? After many days of noticing candles available, and then roofing nails, students figured out how to drip wax to connect nails to rods. When the wax warmed to a certain temperature it loosened its grip on the nail and the nail fell off indicating that the rod was hot at that spot. After much discussion of variables to control such as the number of wax drops, how many nails to use and where to position them, and how to hold the rod end in the burner flame, students were finally able to gather nice sets of data.

Pairs of students typically work at lab the second half of the period and then scurry out. The first task the next day is the reporting of individual lab data to the class. This allows daily practice in the design of data tables and it refreshes the memory of the previous day's work as each lab tells us their findings. Since all data are entered on a grid it is possible to compare numbers as each new set is presented. There are many comments of, "Good, good, that is consistent." Or perhaps, "Oh no! That is crazy data! Truthful but crazy!" When we then plot all the data points on a scattergram graph we can pick out "mainline" trends and outliers. We can speculate over what might have caused a discrepant data point. We talk about confidence in our data, even though I don't quite get to "confidence intervals". We don't have much confidence in widely dispersed data, but we find it interesting because it is truthful. The team reporting the data is recognized as useful and interesting even though we feel they might be wrong.

Now, if we averaged the data, that interesting outlier would be wiped off the map. We would no longer notice that point. Furthermore, all of the clustered data would be slightly shifted. It is easy to understand why students think that averaging data tells lies.

Averaging stops the conversation.

Not only that, it alienates the human spirit.

Don't Let Them Explain

It is not about telling the teacher.
It is about student – student conversation.

Wellington came into class with sheafs of paper. He shyly sat
down with them and began writing more. Yes, he was ready to
take notes on the day's "lecture" (as I call it) but he kept writing
on the side paper at gaps in the discussion. So, I asked him
what he was writing. He said his family had talked about the
current science investigation (how candles burn) last night and
he had talked to friends about it this morning. He was building
an argument for why you could not see train smoke far from the
train. The argument had to do with dispersion of molecules
with no awareness of chemical change of carbon to invisible
carbon dioxide.

What a great line of reasoning! He wanted to talk to me about
it. I said, "Can you tell the class what you are thinking?" As he
discussed his reasoning, classmates responded and elaborated
on his ideas. Finally, at length, the teacher reminded them of
the equation for photosynthesis – and that for burning. An
amazing reversal of terms. The discussion involved engaging,
roaring, extravagant comments. What fun! Really logical,
elegant fun.

Now, if I had let Wellington debrief his ideas to me, he would
have gotten feedback, honed them, refined them, and presented
a rather finalized synopsis to the class. Instead I invited the
class into the presentation. This drew all the students into the

development of the ideas. Right and wrong – they were all engaged.

It seems to me that the purpose of the coming together of a class is not to memorize correct ideas. It is not even to try out steps of learning and correct them incrementally as in programmed learning. The purpose of class is to have roaring debates on substantive stuff.

Teen-age years are fraught with danger these days. Teen-agers spend a lot of time fitting into a groups and experimenting with social activity. They get involved with drugs and deviant behavior. Perhaps what is going on is they are searching for ways to have an interesting life. The human spirit is surging. If school is a place where children swallow words and spew them back at a teacher, where is the soaring human spirit? Spewing and repeating and even explaining are not joy. They show competency.

The task of the teacher is to enthuse students so they seek to struggle to understand. Much of what children learn in school will be considered wrong or useless in a few years. **It is not the knowing of stuff, but the seeking to know stuff that matters.** Teachers will soon pass out of the lives of students. Who then will they talk to? Will the teen-age society they encounter allow them to consider deeply interesting ideas?

I don't let them explain to **me**. I let them form the habit of discussing interesting things with each other.

Reporting Data

We share data at the start of class nearly every day. But how do we do it? What format do we use. Do students get out of their seat and come to the front of the room to speak to the class? Do they at least stand up? No, not at all.

Data should not be a speech. Not a "Fait accompli". Not a "done deal". The purpose of presenting data is not to "report" but to share. We try to make it like a dinner-table conversation. Students can not mediate their own behavior so I act like an arbitrator, enabler, scribe for them. I sit at the overhead projector with students desks clustered in a horseshoe around. I take notes on the overhead as they take them in their notebooks. Whatever I write on the overhead screen, they write down. In addition they are encouraged to write notes to me in the margin of their paper as important ideas occur to them. Each and every one of us (teacher included) makes handwritten tables of the data. I rarely use photocopies in class except to provide illustrations of body parts, etc. I never give worksheets to fill in. The message is, we are working on this together – so we each do it by hand.

In that same mode, students do not come to the front to report. I make a data table on the overhead listing labs 1,2,3,4,5,6,7 down the first column of the paper. Across the top are the titles of data items we decide to share (guided by teacher prompts, of course.) Lab #1 partners negotiate who will verbalize "this" part of their findings and who will present "that". They call out their data points while remaining seated and everyone, including the teacher, enters their information in their tables.

Next, lab #2 does the same. As the data are listed in columns, it
is quickly evident by comparing the number above, whether
data agrees or not. As subsequent labs report, everyone
mumbles aloud about the spread of numbers and trends in the
data. Notice, they have to each write it to digest it. Everyone
graphs it too. There is something about a personally
constructed image that makes a lasting memory.

At the end of the filling out of the data set for all labs in the
class, the teacher writes "Conclusion:" on the overhead.
Students then raise their hands and propose generalizations that
can be made from the assembled data. Although there maybe
individual opinions, usually there is general agreement about
the range of variability of the data and implications of that
finding.

This, then, leads to the next idea related to the concept under
study. Sometimes students pose the next thing we ought to
investigate, but more often the teacher poses a lab which
juxtaposes a new data set to contrast to the last, so new
generalizations arise. Thus after spending several days
gathering data about heating water, and making heating-cooling
curves, I asked if salt water would do the same. There is no
way students would have guessed they were being lead into the
concept of Latent Heat. That was a discovery to be made from
several future labs. For now, they just follow my suggestion of
comparing the two curves.

Informal. Collegial. Students listening to each other as in a
conversation rather than listening to a speech. Students speak
publicly every day but they do so in shared contemplation of the
item under investigation. Learning is genuinely engaging.

Testing! Testing!

Why are we testing? Many times teachers test to evaluate what
students have memorized. Certainly passing an algebra test
records a different level of understanding from passing a
multiplication-table test. That is useful information for
students, parents, and school districts. But is it possible to use
tests in a different way? Could a test promote cognitive
development? Could a test be a tool rather than a judgment?

Today I gave my students a test. They had been combining
drops of pretty colored chemicals to see what colors resulted.
After making (individual) data tables, and discussing results in
class for many days, they began "attacking" each chemical with
a drop or two of a "new" chemical. They worked through
twelve "new" chemicals recording their reaction with the
original yellow, green, clear, blue, and red liquids. Today I
said, "Here is a test to see what you have discovered. It is OK
to use your own notes but not notes others have taken. When
you have filled in as much as you can, turn in your test. It is
OK to make notes in your notebook about which lab results you
need. If your test is turned in, you may take chemicals to lab to
get the results you need to complete the test. Return the
chemicals and get your test back to add more answers."

Strange! What is the function of this exercise? This open-
ended "testing" process motivates students to complete their
laboratory work by making clear to them which tasks are still
undone. It presents enough urgency to motivate them to work
faster and more efficiently. At the same time, it does not create
excessive test anxiety. They have been told that their answers

will become tools for cracking the next code in puzzling
materials. When answers are compared in the following days, it
will not so much be that "I am right" as that "I had uncontrolled
variables (such as dirty eyedroppers) which led me to conclude
a different answer." The test has right answers. The students
will know how well they did. Yet the results are constructive
rather than destructive information.

Testing used this way promotes interesting discussions from
interested students.

Victim, Victim!

Years ago a parent donated 4x8 inch note tablets to the school. We had lots of them. I became accustomed to using them for student comments. Being a distinctive size, I could quickly pull out the group of papers from a pile of other work. At last the supply ran out, but the method of collecting student comments had become so useful to me that I just began cutting 8.5 by 11 inch paper in half. Thus began the famous (or infamous) half-page test. Whenever students see me stand at the counter cutting blank paper in half, they moan or exclaim, "Oh no, a half page test."

Then, eagerly, they say, "Can I be the victim?" There is something enormously engaging about being the main character in the story. This is especially true when everyone knows it is a parody or analogy and the personification doesn't really matter. It is always less fun when I say, "Sorry, no victim today."

Most of the time (but not always) the question I pose probes for their understanding prior to considering an idea in class. Sometimes after several labs I probe to see if any of them can generalize to a new setting. **The purpose of having students write an answer is to have them commit to themselves what they believe at that moment.** Kids are often quick to say, "I knew that'" But did they? By trying to represent their understanding on paper they are forced to recognize their own thinking. Even if they say they know but can't explain, they are aware that their idea is not complete. After the introductory half-page test, several follow-up lab experiences, and class discussions of the implications of their data, students can proudly recognize that they "actually learned something in school today."

I set the scenario. Suzie likes pets. She has a pet rock. The size of a watermelon. Mom makes her keep it outside. The weather gets cold and Suzie is concerned about her pet. She gets a quilt from her bed and takes it outside and covers her pet rock carefully. How long will it take the pet rock to warm up?

So what am I asking? I am really asking, "Do you know anything about metabolism?" But I can't ask it that way or the students will be prompted for the "right" answer.

How about – Bob has a pet fly – named Fred. Fred the fly. Bob decides to give the fly a bath in the bath tub. Just as he is bathing Fred, the doorbell rings. Bob goes to answer the door leaving Fred in the bath for just a minute. When Bob returns: Dead Fred! Why?

Which students will generalize our recent experiments with Adhesion and Cohesion to the ability for the muscles of a small fly to overcome the enormous force of adhesion?

The tests are not graded but students know I have them. Their comments are engraved forever. I never give the papers back nor do I discuss their responses with each person. They know what they said and they remember it for a long time. Class discussion usually erupts immediately and students quickly learn that others agree or disagree. No grading. No humiliation. But lasting memories of reconsideration of their reasoning.

Don't Sit At lab Tables

Don't sit at lab tables. Standing is much safer because you avoid having a lap for hot or dangerous chemicals to land on. Standing allows a student to move fast and efficiently to get the lab task done. Safety is a reason – but not the most important one.

My room has two sections. At one end there is a semicircle of writing desks clustered around my overhead projector. We start every class seated at these tables, where we discuss the previous day's findings and write-up new labs. The remaining three-fourths of the room is lined with counter-height "lab" tables. Each pair of students has their own laboratory space of about eighteen inches by three feet. The space has shoulder high walls like a study carrel. There students display their graphs and set up lab equipment to work on the current experiment. Students do not share the space with others so they can leave their equipment set up. They get the emotional sense that, "This is 'MY' lab."
 "Me. The scientist."   At the completion of each unit of study, we hold a lottery for new lab locations. Thus students "own" a spot for a couple weeks at a time.

This puts high priority on their lab location. Being at lab is fun. If students misbehave (which is rare) or are unsafe (no goggles, for instance), I simply say, "Go sit down." After two or three minutes, I go to the seated student and ask, "Are you ready to go back to lab and be businesslike?" Yes! Yes, they are. Discipline is as easy as that.

Having their own, private lab space makes the science room special. Important things go on here. We use proper laboratory glassware and lab stands. We feel very grown up as we time experiments and collect data. Much of the magic of the room is in differentiation between lecture and lab. The physical space makes it so.

We conduct interesting labs and write down data in one space and we sit and have interesting conversations about that data in another place. Science is great! Everyone loves it.

How Children Invent

Our team is going to the nationals! At the faculty meeting today, the Destination Imagination coach announced that three teams from our school had won awards. One of the teams has been selected to go to the national competition in Tennessee. The teacher described the amazing inventions our students had made to solve the mystery challenge. The team was presented with the task of removing and replacing and assembling things on top of an eight-foot tower. Team members had to remain on the floor yet get the tasks done at a location eight-feet away. Many strategies failed to do the job. One particularly clever invention involved using two different diameters of PVC pipe with a bent coat-hanger on the end of the inside pipe. By retracting the inner pipe, the coat-hanger "claw" could be made to close around an object 8' away. Clever! Everyone felt empowered by their own original invention. Adults were amazed. They had not "taught" students anything but had given them free rein.

No one on the team or in the faculty audience seemed to remember that two years ago we had "primed the pump". Our all-school engineering event had been on the theme; "Lend me a hand." We had invented various prostheses devices. One had been a 4-wheel match-box "go"machine loaded with a drinking straw – bobby pin contraption. By retracting the bent bobby pin into the straw, we had been able to grab a tiny hand. The go-cart sped across the floor crashing into the end wall. The crash pushed an interior coffee stir-stick forward in the drinking straw, forcing the bobby-pin out the end thus letting it spring open and drop the model hand. They had "Lent a hand".

Without being aware of the origin of their cleverness, the Destination Imagination team had built on their prior knowledge. So, how do we learn? Where do wonderful ideas come from? Our various experiences in life are tucked away in unknown places in our brain to be revisited and elaborated as we face new challenges.

Learning and invention are amazing!

Experience matters!

Don't Grade Tests

Don't grade tests. Celebrate them. At least sometimes.

After doing labs related to vascular flow and lymph activity, we spent ten days learning about the heart. We began the first day with mnemonic strategies. Upsie-downsie; Superior Vena Cava, Inferior Vena Cava. We drew and labeled as regular medical illustrations do. It took us about four class periods to draw and label the entire heart and its primary tubes. We started each period making a progressively more detailed sketch. I drew it on the overhead projector and students copied it in their notes. The second half of each period was some light-hearted, novel experiment related to the sketch we had just made.

After sketching and adding detail each day for several days, the students knew the structure pretty well. We talked about Atherosclerosis and Coronary Artery disease, but being rather young, students were more interested in "birth defects". At birth the baby's environment changes drastically. Oxygen had been delivered via the Umbilical Cord but suddenly the lungs had to take over. What changes in the heart were required to do this? The infant heart had only a minute or two to make the changes, so how did this miracle happen? We drew the Foramen Ovale and the Ductus Arteriosus. We examined pictures showing heart structures which were not able to change blood flow properly. We did some labs with occlusion devices and considered what a doctor had to worry about as he tried to save the life of the child.

Finally, after ten days of study in class, (with no homework at all) students were asked to make a "Museum quality" sketch of the heart and surgical procedures. No grades would be given. We would simply celebrate what they had learned and drawn. The total words listed would be mentioned on interim report cards and the sketches would be given to parents at conference time to be taken home and perhaps framed. Student journals, set on tables across the room, could be consulted from time to time as the sketch was developed, but notes could not be taken from the table to the drawing station.

So, how did they do? Scanning down the grade-book, I find these young students were able to draw and label 105, 142, and even 221 terms! Most students had between 80 and 110 terms. The sketches accompanying the words were amazing – as good as college level medical books. Now, there is no way I could have asked my students to learn this much. Students of this age normally learn about twelve terms related to the heart. **By generating interest rather than requiring achievement, my students excelled beyond my wildest dreams. Real learning is not a top-down judgmental affair. It comes from the heart and soul of the learner.**

Tests don't always need to be graded – sometimes they need to be celebrated!

I'm Going To Turn On The Light

Students come in the room and are seated. They take out their notebooks and perhaps head their paper. They talk to their neighbor and eat a bite of snack. After a minute or two, I walk to the center of the semicircle of student desks which cluster around the overhead projector. I insert a new audio tape in the reorder, and say, "I'm going to turn on the light." This is the cue to quiet down and focus attention on the lesson for the day.

I write, "Lecture" in the center of the overhead projector transparency and underneath write a question or statement. Most of the time I reference the activity they were doing at lab as the period ended the day before. Making a data table for class data not only causes them to review their notes and speak publicly to a group, but it refreshed their memory about what we are doing and why we are doing it.

Today we had an assembly. Several class periods were missed. I try to keep the several sections of classes for a given grade doing the same task. Common activity promotes lunchtime discussion of, "What result data did you get in lab today?" It promotes spontaneous, interesting conversations. When students all work on different things, no such discussions are in evidence.

So, with two sections missing class, I was faced with what to do with the third. Since they were working on a long task gathering data about twelve different chemicals, one at a time, I knew students had a lot remaining to do. When they came in the room, I said, "No notes today. Just go to lab." Well, that was a surprise. That almost never happens.

After about five minutes, three different lab pairs had come to me saying they didn't know what to do. I said, "Well, yesterday we wrote down two tasks. 1. Finish the six magic-paper color codes. 2. Choose the most informative code and see what each of the twelve chemicals does to it." Still they didn't quite know how to proceed.

I called out, "OK, everyone come over to lecture for a minute." I turned on the overhead projector light and asked for class input for one set of data. Lab 1, lab 2, lab 3, and so on – until we had a data table reporting information from each lab. Then I said, "Now add to this where you left off yesterday." Students returned to lab and busily went to work.

That reveals something interesting. Students need multiple modalities. They need to write, speak, listen and reflect. They need structure in resurrecting their memories of the task at hand. They benefit from a routine of how we get focused on the topic of the day. They need a certain type of action from me to frame their expectations of what to do in the science room. They love science and they feel free here, but they can't do it on their own.

Class structure matters. A lot.

It Takes Three Times

We wanted to time rod heating. How do we do that? Well, we can put drops of wax in roofing nails attached and attach them to rods. So we go to lab and try it.

The next class session we report data showing that our nails fell off. OK, how many drops of wax works. Are there ways to get the drops similar. We decide to control the variable of drop number by using between 3 and 4 drops per nail. We go to lab.

Reporting the next day, we have wildly different nail-drop times even though we are using nearly identical rods. How can that be? Where did you place the nails? Lab 1? Lab 2? Etc. Well, we want to know when the entire rod gets hot, so we probably should all agree to put a nail at the far end. But can we detect heat coming down the rod? How should we space the nails? We agree to wax them on every two centimeters in from the far end – and to use six nails.

Hummm. We make a class data table and each lab reports nail drop time. Our times are starting to follow a pattern. The variability between labs for nail #3 is very small. Can we do better? What other uncontrolled variables could be affecting our data? What do we need to all agree upon? Well, we need to make sure we heat the entire rod. No "foot over the starting line." The rod end needs to be IN the flame. The flame needs to continuously curl around the rod end. And then, the rod needs to be held horizontally. Tipping it up may change the data. What else? What else? We are getting closer.

It does not matter which experiment we are doing. If our intention is to teach the "process" of science, we need to set up experiences where students discover messy data. They then naturally seek to control variables and "clean up" experiments. They come to understand why finding the cure for Cancer takes so long. Real science doesn't know exactly where it is going and it doesn't know how to get there. Real science understands that truthful, repeatable data is important and may actually be significant.

For this age child, it usually takes three days to understand the question, control enough variables and compare enough data with other lab teams in the room to get satisfaction out of a data set. Students feel they are making their own original discoveries.

 And, in a way, they are.

New Seats

Just come in and sit anywhere. We will decide on the seating arrangement in a minute.

The teacher draws the floor-plan of the classroom on the overhead projector screen. Hall door is written in one corner to orient the map. With a partially straightened paper clip, a seat is pointed to. The teacher covers a tablet with fifty or sixty single digit random numbers.

The teacher points to a student. "What are the numbers between?" "What? What are you taking about?" The teacher asks another student, "What are the numbers between?" Finally a student replies, "Thirteen to fifty-six." The teacher closely holds the paper containing the random numbers and secretly circles a number.

So, "Who wants this seat?" Hands go up. "OK, write down your guess – a number in that range." "What was the range?"

"Ask the person who gave the range."

Students are sitting in a semicircle making it easy for the teacher to point to each child interested in that seat. It seems fair. No jumping around. The teacher points from child to child sequentially across the room. "Not written down?" "Sorry, you are disqualified."

Students are too clever. They know how to play number games. If they hear a number spoken, they choose the next larger number to cover all choices larger than that. No. Students must each write down their number choice before they hear the choice made by anyone else. The strategy then becomes, "what is the likelihood that the teacher's hidden choice is low in the range – or high?"

When all the students wanting that seat have stated their number choice, the teacher unfolds the random number paper and reads off her choice. The student with the closest number gets the seat.

A new seat is pointed to. A new student is pointed to, to select the range. The teacher secretly circles a new number in the new range. Students write down their choices and a new seat is awarded. So it goes until everyone has a seat.

It is amazing to watch the process. No directions were given by the teacher. Every child is engaged. If some child is so confused as to be unhappy or upset, other students quickly intercede and try to explain what is going on. Not only do students get new perspectives within the classroom every time we do this, but they come to understand that this is a place where you have to pay attention and be clever. Things are not explained. There are rules, but you have to figure them out. As time goes on, kids get more clever. In December, Julian was asked to choose numbers. "Between -2 and - 2.5." The kids dive in. There are lots of numbers in that range. Math is alive and useful.

A student notes, "Science is interesting because you have to figure it out on your own."

Ready To Light, Doc "O"

Today students were making "museums" showing physical change and chemical change. They each had an 11-inch by 17-inch foil cooking tray. Inside they placed an 11x17 piece of computer paper. On the paper they drew a large T and labeled the columns "Chemical Change" and "Physical Change".

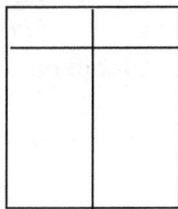

They were to choose an object, such as a stick, pencil, or post-it note and divide it into two parts. The first part was broken showing physical change and taped to the physical change side of the "museum." Now they were ready to make a chemical change on the other part. They were to light it on fire without burning down the school. How do we do this?

In order to have freedom, you have to have boundaries. To be free to drive a car, you have to recognize stop signs and stop lights. It is helpful to have lines painted on the road so you are confident about where to move and what to avoid. So it is with science lab.

First of all, "You are meat – you cook." We need a "Fire extinguisher beaker." This beaker, half full of water (not totally full), has three uses. First, it is where we place lit matches. We do not blow out matches. We do not bring flame near our face.

We simply light the match, light the burner, and drop the
flaming match into the beaker containing water. Secondly, you
are meat – you cook. If you ever feel like your finger is burned,
put it in water right away – then say "Ow". That is how fast
you put your finger in. Quickly. You are like cookies taken out
of the oven. You have to cool. If you cool a burned spot
quickly enough, you won't get a blister. The third use of a
'"fire extinguisher beaker" is to pour on lab fires.

The matches are kept in a "match-Petri". This is a Petri dish
with a strike-plate from a box of kitchen matches taped on the
lid. Matches are available on a counter and students place them
in the Petri. They only choose a limited number of matches.
"Don't take more than the size of a bonfire you are willing to
put out." Three or four is a good number. Matches are never
carried singly around the room. The only place you can carry
them is in a covered match-Petri. Bring the covered, empty
match-Petri to the supply table, get the matches you want, then
take the covered match-Petri to lab.

So now, we are ready to strike a match. No one ever strikes a
match in the science room without Doc"O" standing next to
them. Your sample is in a metal test tube holder, the burner is
in a small foil tray on your lab. (This is not the museum tray)
The fire-extinguisher beaker is sitting at lab next to the tray. So
you loudly call out, "Ready to light, Doc 'O' ."   Then you
wait. Quietly, patiently, until she gets around to you.

You light – snuff , and you don't burn the school down!

Go To Lab

I am told that in colleges of Engineering, labs occur perhaps once per week. They happen in a different room than the lecture hall and they are on loosely related topics but not necessarily the topics of the day. When I visited China I learned that labs may be offered in a separate course in the summer with lecture classes occurring during the main school year. Does that matter?

Science information is often "presented" as material to be memorized and tested. Science labs are guided by printed instructions and forms with spaces to fill in data. The "right" answer is required to get a good grade even if the lab equipment produced marginal results. It is more important to be correct than to be truthful.

No photocopies are used in my room. No student tape recorders either. I write by hand on the overhead projector screen and students copy this into their notes. When students speak a reasoned response into my tape recorder (I try to "catch" them being intelligent), I paraphrase it in my writing and the students do the same.

So, we discuss. We take data. We together decide conclusions and contemplate why data agree or disagree between labs. We discuss uncontrolled variables and confidence in trends in the data. From this we draw the next task to perform at lab.

We need each other. Our lab equipment is rough and our data is imprecise. We write our lab instructions together as we decide what steps to do. Our write-up is strongly guided by teacher input since students do not know the discovery to be made. But we do know the activity is OURS, not handed to us in some photocopied form. Only by comparing six to eight sets of data can we gain confidence about whether ours is close to being "correct" or not. We can say, "In a perfect world we might get…" The most important thing is that we are truthful. Everyone's data point is honored for its truth in reporting. No fabrication of data. **"Wrong" answers are interesting!** What are the uncontrolled variables? How much do they affect the outcome? We don't get a better grade for getting perfect data. We do not rewrite the lab in "final" form, but leave it with uncertainty and tentative results. We are learning to live with ambiguity. (We are never certain we will get to school safely, but we still drive cars.) As a matter of fact, **perfect data are suspect**. When graphing the weight of increasing numbers of milliliter "pots" of water in test tubes, we recognize that, "A straight line is a wonderful thing," but if our line is too perfect, we wonder why. Wouldn't some drops of water adhere to the side of the test tube from time to time? Wouldn't our meniscus be just a bit wrong once-in-a-while?

Yes, we love the "right" answer, but we value the truthful answer.

Science is not perfect. It is built on the seeking of truth in a complex world with people's imperfect perception.

Truthfulness vs. correctness. We seek the second but we value the first even more.

Hot-Rod Coming Through!

The students had been challenged to "run hot-rod races." Oh, you mean HOT rod races!
Yes, who can make a rod hot fastest? How can we know when the rod is hot? How hot does it have to be? How can we know if it is hot enough to be a "win?" It must be VERY hot. How do we handle dangerously hot items in the classroom? How do we organize the room so students can do such dangerous things as light a match, light a candle, light a burner? How can we let students operate on the edge of their comfort zone, yet keep them from having accidents?

First, labs need to be separate cubicles, not side-by-side counters.

The supply table is at the end of the room. Students get supplies and take them to their lab work stations where they can peer over study-carrel like walls to see others working, yet not jostle elbow to elbow. Students working in pairs stand side by side between the tables. The teacher can quickly move from lab to lab, attending as needed, or can observe unobtrusively from across the room. Students feel they have great freedom, yet they are constantly monitored.

With metal trays on each lab table, each pair of students has a burner to light - and then extinguish. Students work intently, bending over their lab work area to do the job. They do not pay attention to the location of flame. If the burner is lit while they are dropping wax from a lit candle, accidents may happen. Leaning over brings their hair close to the burner flame. So, only one flame at lab at a time. Students light the candle (Ready to light, Doc "O".) and use it to drop wax on nail heads. They blow out the candle. When they are ready to heat the rod they must call out, "Ready to light, Doc 'O' " again so they have close supervision while lighting the burner. This gives lots of lighting practice.

Notice, I never light a match for them. Students hands quake as they strike a match for the first time. I will reach over to grab their wrist and twist their hand so it is lower than the match flame, but I will not light a match for them.

Now, they have been heating rods. What do they do with them? How do you get rid of a hot branding-iron? The hot rods have to be placed on a metal tray on the supply counter. Students need to carry the hot object to the far end of the room. Other students need to get out of the way. On the highway, ambulances turn on sirens when they need to get through traffic. Cars move aside. So, we become ambulances. "Hot rod coming through!" "Hot rod coming through!" The students step out of the way when they hear the alarm called out. The student with the hot rod (held in a wooden clothespin) carefully (and somewhat slowly – watching others clear a path) brings the hot rod to the supply counter and drops it there. Whew! A danger delivered safely!

No Nylon Coats At Lab

Nylon melts. Nylon drips as it melts. The drips stick to things. Hot drops of nylon sticking to skin burns and burns in an annoying way. "No nylon coats at lab."

No scarves or floppy coats either. We need to wear clothes which do not interfere with the dangerous things we do at lab. Hair styles change. We often love to wear long braids or curls. Bring a "scrunchie" or get a thick rubber band from Doc "O". Hair needs to be held back so it does not fall forward to drag along our lab materials.

Goggles seem very grown-up. We each choose our own and put our name on tape on their side. Goggles hang at lab from little hooks made from bent coat-hangers. They are always ready for use. Goggles must be worn when we light alcohol burners. (State regulation, you know). Glass alcohol-burners could crack and "explode". We don't have to wear goggles when we light candles, but perhaps they are useful in keeping the hair back.

Shoes in the science room. You must wear shoes in the science room.   We try to sweep up broken glass, but you never know when we missed a bit. Wear long pants. Don't wear sandals. Lab supplies drop from time to time. It is best if we protect our legs and feet from falling things.

Now, wipe your feet. After working with candles for three or four days, the science room becomes like an ice-rink. Even when students try to keep wax on the lab table, occasional drips land on the floor. We step on it and smooth the wax out over the room floor. Soon, every step we take is slippery. Students start running and sliding across the room. No! Not in the science room! That is not safe.   Sweeping the room doesn't solve the problem. So we put a soaking wet bath towel on the floor. As students come and go, they are to pause at the towel and wipe their shoes. In this way, some of the wax is scraped from the sole of their shoe. They step off the towel with better traction.

Pretty funny. It is pretty funny watching students try to wipe their feet. They step on the towel and slide it back and forth. The towel, that is. They wipe the towel on the floor but they do not wipe their soles with respect to the towel. Students have to be instructed how to grab the towel with one foot and scrape the sole of the other shoe over the towel away from that spot. Wiping feet takes skill!

So many little nuances to safety. The teacher needs to be observant to every little childish action which might lead to a hurtful event. Rules are created whenever a new concern is noticed. The students respond appropriately to new rules because they know that rules allow freedom. If you can't be safe: sit down.

When everything is set up and safe, call,
"Ready to light, Doc "O" ."

Laminate Graphs

Students generate long data tables. They share data day after day in class. After a week or so of heating water from ice to boiling and taking data every fifteen seconds , they have enough data to make a graph. In the process they discover the perplexing issue that, although heat is continually put in, the temperature stops rising. Heat and Temperature are not the same! That is odd. At first they thought the thermometer was broken. Something must be wrong! But finally, students came to realize that the world was telling them something amazing. In fact, Heat and Temperature do not always give us the same information. After more experiment we come to recognize that heat can be hidden. Latent Heat! What an amazing concept.

Graphs are carefully crafted and turned in to be laminated. Freshly laminated graphs are proudly taped to the walls of student labs. They love them. They "own" the data. When visitors come in, students can point to their graphs and explain what we have been doing in science. Students develop an emotional attachment to their graphs. They are meaningful and beautiful in all their simple mathematical glory. No decorations on the edges – just the careful presentation of hard-won data.

Graphs remain at lab for many months. Students choose new labs every month or so and the graphs move with them. Their graphs and their goggles establish the territory as "theirs" for now. But more than that, a glance at the graph from day to day, reminds students of the concepts they have learned. If we put them away in a file, the idea of Latent Heat may become a distant memory, or indeed it may be forgotten.

But science is more than just learning concepts. The sequence of units is crafted so that ideas reemerge in surprising ways. We start the year with Latent Heat. We spend long enough on it to gain some understanding. We even make "Old Fashioned ice cream" along the way. This only works because we have discovered that the "low plateau" of salt water, fresh water and sugar water are quite different. Amazing interactions gather heat from the cream to freeze it.

But at the end of the school year we study Earth science. At last we ask the question, "Why does the Mantle churn? Where does the heat come from?" I remember Erik's exuberance when he connected old fashioned ice cream with the core of the Earth. He was radiant as he explained that the core is solidifying, so it must be getting rid of heat. The heat must move outward to give off heat to the mantle. The mantle, heated from below (like a stove) generates convection cells which rise and move the continents. Stunned silence! "You set us up!" "Do you mean you have been planning for this moment from the start of the year?" "Our teacher tricked us again!" But what a delightful, intellectually powerful trick. "We got it." "Latent Heat, again!"

Yes, a lovely idea. But a mind-boggling idea when you make the connection yourself.

Figure It Out – Or ...

Figure it out, or ... I will tell you! "Oh, no!" "SHE IS GOING TO TELL US! Now, that is rather odd. Aren't teachers supposed to "tell" students? I have noticed parents in previous schools in which I taught, complaining that their child could not do the math homework because the teacher did not explain how to do the problems well enough. Teachers tell, students memorize. Books tell, students memorize.

I have found the BSCS (Biological Science Curriculum Project) interesting over the years. A few decades ago they produced several versions of High School Biology books. Green version, Blue version, Yellow version and Black version. These were used enthusiastically by teachers for a while but students complained or didn't sign up for classes. The way I view it, is that the books were too "wordy" for the type of students who want to figure things out, and they asked for too much reasoning for those willing to read all of that. A wonderful project! But it didn't last. I still love some of their clever questions, but I have to pose them myself without using texts. These clever questions engage students best when the students think they have come upon the idea on their own. Books don't work for me in class. Although I, personally, read all the time. I love books.

No, the task in the classroom is to cause kids to want to go to lab. You can't ask, "What do you want to study?" That request doesn't make it fun. The teacher has to contrive a question in a way that grabs student attention. I often start the year with a "Half-page test" asking, "How hot can you heat water?"

Students typically say, "To 100°" or "To infinity." Either way they are wrong. (Because they haven't considered elevation.) Even those who say, "I have no idea" know that they do not know. I can then, with them, design a lab asking, "How hot can you heat water?" After taking data every 15 seconds for a half hour, students begin to notice what some of the questions are. They make beautiful heating-cooling curves, but they recognize that the curves just lead to more questions. Most students want the understanding to come from analysis of data from class members, rather than from the teacher "telling" them.

So, does this method "Leave some children behind?" My view is that, with this method, every child has been given an authentic opportunity to make discoveries from their own data. If they do not notice trends, other students may tell them. Being told is no worse than reading the answer in a book or being told by a teacher, as in a typical science class. No child is "left out". No child ends the unit without learning the idea.

There is an added advantage. The child who "doesn't get it" watches another child make the connection. Students see it is possible for people of their age to make connections. Rather than thinking that they are supposed to be passive and be told, the "quick" students provide models for the others. The excitement gives everyone "permission" to try to figure it out.

Not every child is going to discover the connections. That is OK. It may not seem fair, but the alternative is for everyone to read a book or be told by a teacher.
Perhaps we need to reconsider our concept of fairness.

Teachers Say:

"I have known Dr. Olson... as both a teacher and a professional colleague. My initial contact with her was as a participant in one of her workshops. Her conception of the teaching of science and mathematics is unique, powerful, and unquestionably worthy of national recognition. Focusing on the students' logical skills and personal ingenuity, Dr. Olson's approach reflects an amazingly enlightened understanding of the essence of scientific and technological progress. I was so profoundly moved by the ideas she presented that it became an immediate ambition of mine to learn as much as possible from her and possibly one day work with her. The thoughts she set in motion helped me formulate and lead a highly successful public school program for gifted children for 8 years ... one of the most extraordinary teachers I have ever known."

Larry Guldberg, Lakeside School

"This woman has absolutely the best grasp of the ramifications of gifted education that I have ever seen. Her experience, powers of observation and ability to correlate this information are phenomenal. This praise may seem overblown but I have read most of the more well-known educators of the gifted and there is no doubt in my mind that Dr. Olson will be recognized (if not already) as the top in her field in the country. Amazing."

Teacher, Seattle University student

"Dr. Olson has, once again, presented a course which has improved my professional competence, enthusiasm and personal awareness. She has a gifted ability to provide theory, research, practical ideas, resources and professional personal support. I have never been so intellectually challenged and motivated, while still leaving the course with many functional materials to implement. I have truly grown as a professional because of her course.

Teacher, Seattle University student

"This was an excellent course – ought to be given to all education majors! Great job – excellent teacher."

Teacher, Seattle University student

Administrators Say:

"Meredith is by far the most brilliant and outstanding teacher that I have ever known. She has exceptional competency in the disciplines of science and mathematics. Her sensitivity to the children's cognitive style has allowed her to develop teaching methodology that addresses the individual needs of each student. Our students have demonstrated competence to the point that receiving secondary schools have needed to upgrade both their science and math programs. Our alumni have repeatedly referred to Dr. Olson as the most influential teacher in their memory. As a direct result of her exceptional ability our school has gained the reputation of having the most outstanding science department of any private school in the Northwest. Many parents select SCDS as their private school of choice because Dr. Olson is on the staff."

Lucile Beckman, Former Principal, Seattle Country Day

"Dr. Olson came to our district as a workshop leader in an experimental teacher training program. She spent one week with us here, teaching gifted children in the morning in a summer seminar, and working with teachers in the afternoon, explaining what she had done with the students and elaborating on methods for teaching this type of student. Her impact was so significant that the teachers have repeatedly requested that she come back for a longer period of time. Dr. Olson's extraordinary skills as a teacher are matched by her extraordinary levels of commitment to the task. She gave personal attention to the needs of students and to the needs of teachers, recognizing their limitations but able to adjust her instruction to make the learning not only effective but tremendously exciting to students and teachers alike."

Marjory Ward, Lexington County School District 5

Parents Say:

"We just wanted to take a minute to thank you for a wonderful year of science. Anna's enthusiasm for what she was learning in science was contagious and made up a big part of our dinner time family discussions. I know you hear this time and time again but you really have the magic for igniting curiosity in young students. Coming from a family of scientists it is a pleasure to experience Anna's love of the subject this year. Have a wonderful summer."   TT & PT

"Thanks for everything you do for the students. Your class is a big reason Joshua enjoys going to school. I hope you have a wonderful Thanksgiving holiday !Sincerely,"   KG

"I wanted you to know that on at least a weekly basis he tells me some fact about the physical world or how something functions. I used to ask him where he learned that. I am now beginning to say "Did Dr. O. teach you that?" and he says with a smile "of course, who else?". Your presence in Skyler's development is huge and I know the intellectual curiosity you support will strengthen his ability to think independently throughout the rest of his life. Thank you so very much for your contribution to these years of learning for him. Please know that Skyler holds you with genuine respect and fondness. Warmest regards." CC

"I am so glad that both Brice and Eleanor had you as a science teacher. What you do with this age group is both unique and extraordinary. You have shaped the way they think more than any other teacher. I am sure that they will be swapping Doc "O" stories seventy years from now. Take Care." JR

"When every student I know says that science is their favorite subject, it means that you have achieved what most science teachers only dream of.... You have captured the imaginations of these kids and made science fun!   Thank You." MC

"Thank you for the enthusiasm you bring to the classroom. *Jason has really developed a love of science during these two years. You have given him that gift and we are very grateful. Wishing you the best!"* LW &KW

"I'm sure we're not the only parents who say you're one of the reasons we wanted our son to attend SCDS. And now that he is in your class it is all the more true! Every single day, Ben jumps in the car after school and says science was the "most awesome" part of his day. Thank you for showing him the creative fun of thinking like a scientist." LO & MO

"What an amazing year of science! Madison loved every second of it – being challenged, having to work towards an answer that wasn't always apparent, and the independence and trust. She took a tremendous amount of pride in her work and the experiments were a great source of material for our dinner conversations. Thank you for being such a lovely consistent and inspirational role model for Madison. You may be one of the few teachers who has ever really "gotten" her. Warm Regards." KO & DO

" Molly and I were recently talking about the move to her next school. What she was looking forward to and nervous about. She stopped talking (which, as I am sure you know is notable!) and she became a little misty. She said she was thinking how much she was going to miss you and your class. We just wanted to let you know what an amazing, enriching, mind-broadening experience Molly has enjoyed these last two years. I don't know if she will have a future in science but she has certainly become "richer" because of your science classes. "Doc O" is often quoted in our home. (I do think liberties are occasionally taken with the quotes!) Our spring break was spent with friends in the kitchen concocting and blowing stuff up! Our entire family benefited from Molly's two years in your class. Molly's younger sister is looking forward to science with you in the not too distant future. Thank you does not seem quite enough for a one-of-a-kind and remarkable woman. We hope you enjoy your summer and we look forward to our next child's science with you. Thank you." JM & SM

"My son Aaron thinks that you are the "best teacher in the world." He talks about science class with boundless enthusiasm and we hear "Doc O says.." all the time. He is sure the world would end if he missed your class. He wants to be a scientist just like you when he grows up. Thanks for making such a difference in his life. You are the teacher he will always remember. Sincerely" SLS

"Yours is the one class that Erin regularly comes home wanting to tell us about. She has so enjoyed science class this year. Thank you for being so inspiring! Have an enjoyable summer. Warmest regards." MC & YPC

"Thanks to you Zoey is on fire with excitement about science and anything related to scientific inquiry. She looked forward to your class every day with unbridled enthusiasm. Often Zoey's conversation on the way home from school or during dinner focused on what she'd done or what you'd said in science class. You have instilled in Zoey an appreciation of the scientific approach to observing, analyzing and solving new questions and problems." RS

"Thank you for such an <u>amazing</u> science experience for Natalie the past two years. We have all completely enjoyed your teaching style and mentorship." DF & MF

Kids Say:

I think the atmosphere here is really different from any other class. We write the exact same things down as everybody else in class. Yet, we don't get bored at all. This type of environment with lectures, labs, and no textbooks or anything, gives us a sense of freedom. I walk into this classroom and just feel immediately excited about the period coming ahead, no matter how the last one went. We don't ever really have big tests of random facts that we don't even understand. Instead of that, we focus on things that you would probably see in middle school textbooks, i.e. Carbon/Hydrogen, the human gut, etc. This class is just, I don't know, different, spectacular, maybe? This is an amazing class, taught by an amazing teacher, overall. GS

Science is a very different class. Of this statement I am entirely sure that what I say is true. It is a fun class for many reasons: We have lab, in which we do experiments. We write down a lecture, which is oddly fun for a reason I cannot realize. And we have Doc "O", the most intriguing teacher ever known. DG

While in L.A. we have to listen and shut up. You don't make us do that. Also you let us burn things which is really, really, really, infinity fun. Plus I like the smell of burned things. It reminds me that if I don't like a tree, I can always chuck it in the flames and watch it burn. There's something magical about watching things shrivel into dust. I also like how you make us

take notes. The 1/2 page tests are fun. The other classes have boring tests about Beethoven or the war of 1812. I also like the fact that you let us have class discussions. Those are so much fun. I also like the names for things like, "Fattie and Skinny." I also love the engineering event. It is so much fun to say, "My rig went farther!" At my old school we didn't have science class. All we did was learn about bugs and metamorphosis. Then again, we lived in Arizona. It sucked! It was way too hot. The school was under-funded and we didn't have science.
CF

I think it is ten times more fun than any other class because it has labs and you get to do all this crazy stuff like light burners and try to melt candles, and try to catch "naked bodies." You also get to write up lecture.
EF

You don't tell us, you help us.
Really interesting.
You make up stories to help with learning.
WF

We get to use things other science teachers don't let us use. We get to try and figure it out on our own. When I was in public school we did it the most boring way ever.
CE

You are not boring at all. You make class fun!
OZ

We write down lectures. We don't fill in blanks on test papers. We have labs and it is something we are willing to talk about to our friends and family. We also learn a lot in fun ways.
HR

I love coming to science. Science is unusual because <u>we</u> figure it out over a course of time, where as in other classes we wait about 2 minutes before we are told.
ED

We have lectures so we know what we need to do. We get up from seats and do the lab. In Humanities we sit and listen to teachers and do work sheets. We don't get up and wonder around. In science we do which is good. This one is unique. I like this science class because it is different.
EH

A Neighbor Says:

I've written earlier this year about my landlady, whom I judge to be the archetype of Ms Frizzle of the Ms Frizzle and the Magic School Bus series. Of the hundred or so times I've hitched a ride to the University with her, on no occasion has our 20 minutes together been lacking in stimulating ideas, unanticipated revelations, and thoroughly delight-full discourse. It is one among myriad blessings of my sabbatical to have landed in Seattle within her sphere. She is responsible for an annual engineering challenge at the school where she teaches, and this year's is inspired by news of the Archimedes Palimpsest having again broken into the public consciousness.

Stuart Weibel
Library & Internet standards wonk.
Online Computer Library Center

Weibel Lines
Ruminations on Libraries, Internet standards, and stuff that comes to mind.

About the Author

Meredith Olson Ph.D.

Dr. Meredith Olson, known affectionately as Doc "O" to her students, has taught elementary, middle school and high school math and science in Seattle for more than 50 years. Her primary goal is in improvement of pre-college engineering education. By going to lab to work on contraptions every day, her students come to understand properties of the mechanical world.

"It has been a long and interesting trip. Studying some metallurgy in grad school. Evening classes. After a full day of high school teaching. Consulting for JPL as the Mars Pathfinder Educator. Weekends. Working in the summer with UNESCO in Zimbabwe, Kenya, and Uganda. Teaching dozens of weekend and week-long summer teacher workshops in South Carolina and Montana. Being a consultant and curriculum designer for Health and Physiology education in Washington, Oregon, Idaho, Montana, and Alaska. Being a summer adjunct University instructor for more than 20 years in Seattle, Idaho and Montana. Teaching teachers. Teaching students every day, every year for 57 years. Observing how learning happens. Becoming aware when real learning isn't happening. When it is just "show." When it is just teacher–pleasing to get a grade. To get a credit. To get a university degree."

See Dr. Olson's open letter outlining her philosophy of lesson design, available on the JPL website - *Exploring Preface* pp 11-13 http://mars.jpl.nasa.gov/education/modules/GS/GS07-19_preface.pdf

"I think right from the start, the classroom dynamic taught me to listen. After 57 years of daily classroom teaching I may be getting better. I may be better at listening to students."

Dr. Olson believes that children must construct their own understanding from active design and assemblage of contraptions. By testing, failing, remodeling, and trying again, we come to **see** the structure when we look. By carefully examining materials we have, we may perceive how to use them in new and unexpected ways. Children begin to understand the engineering process. Besides, it is fun!